Worms

By Theresa Greenaway

Photographs by Chris Fairclough

RSVP
RAINTREE
STECK-VAUGHN
PUBLISHERS
A Steck-Vaughn Company

Austin, Texas

Published by Raintree Steck-Vaughn Publishers, an imprint of Steck-Vaughn Company.

Acknowledgments
Project Editors: Gianna Williams, Kathy DeVico
Project Manager: Joyce Spicer
Illustrated by Jim Chanell and Stefan Chabluk
Designed by Ian Winton

Planned and produced by Discovery Books

Library of Congress Cataloging-in-Publication Data
Greenaway, Theresa, 1947–
Worms/by Theresa Greenaway; photographs by Chris Fairclough.
p. cm. — (Minipets)
Includes bibliographical references (p. 30) and index.
Summary: Provides information on the identification, life cycle, and
habitats of earthworms and other kinds of worms, as well as on how
to collect and care for them as pets.
ISBN 0-8172-5588-5 (hardcover)
ISBN 0-8172-4208-2 (softcover)
1. Worms as pets — Juvenile literature. 2. Earthworms as pets — Juvenile
literature. 3. Worms — Juvenile literature. 4. Earthworms — Juvenile literature.
[1. Earthworms as pets. 2. Worms as pets. 3. Earthworms. 4. Worms. 5. Pets.]
I. Fairclough, Chris, ill. II. Title. III. Series: Greenaway, Theresa, 1947– Minipets.
SF459.W66G074   1999
639'. 75 — dc21   98-34075
CIP   AC

1 2 3 4 5 6 7 8 9 0 LB 02 01 00 99
Printed and bound in the United States of America.

Words explained in the glossary appear in **bold** the first time they are used in the text.

Contents

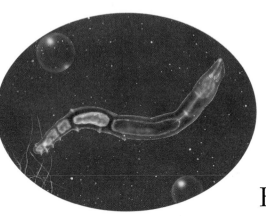

Keeping Worms

Worms were probably some of the first creatures that you met in your backyard. Because they move so slowly, you can take a good look at them before they slither away.

Not many people know exactly how worms live. Keeping worms is a good way of finding out more about these useful animals.

A worm is usually no more than a few inches long. It is shaped like a tube. A worm's body is divided into rings called **segments**.

◄ Always wash your hands after handling worms.

There are about 150 segments in most earthworms. Worms have no eyes and no antennae, but they do have a mouth. Watch to see in which direction a worm moves. Then you will know which is the front end.

▼ Each earthworm segment has bristles that help the earthworm move quickly when in danger.

Worms give off slimy **mucus** all over their skin. This keeps them from becoming too dry. It also helps them slide through the soil.

Worms will soon dry up and die if exposed to warm sunshine. Their skin is sensitive to light, touch, and the presence of moisture or chemicals in the soil.

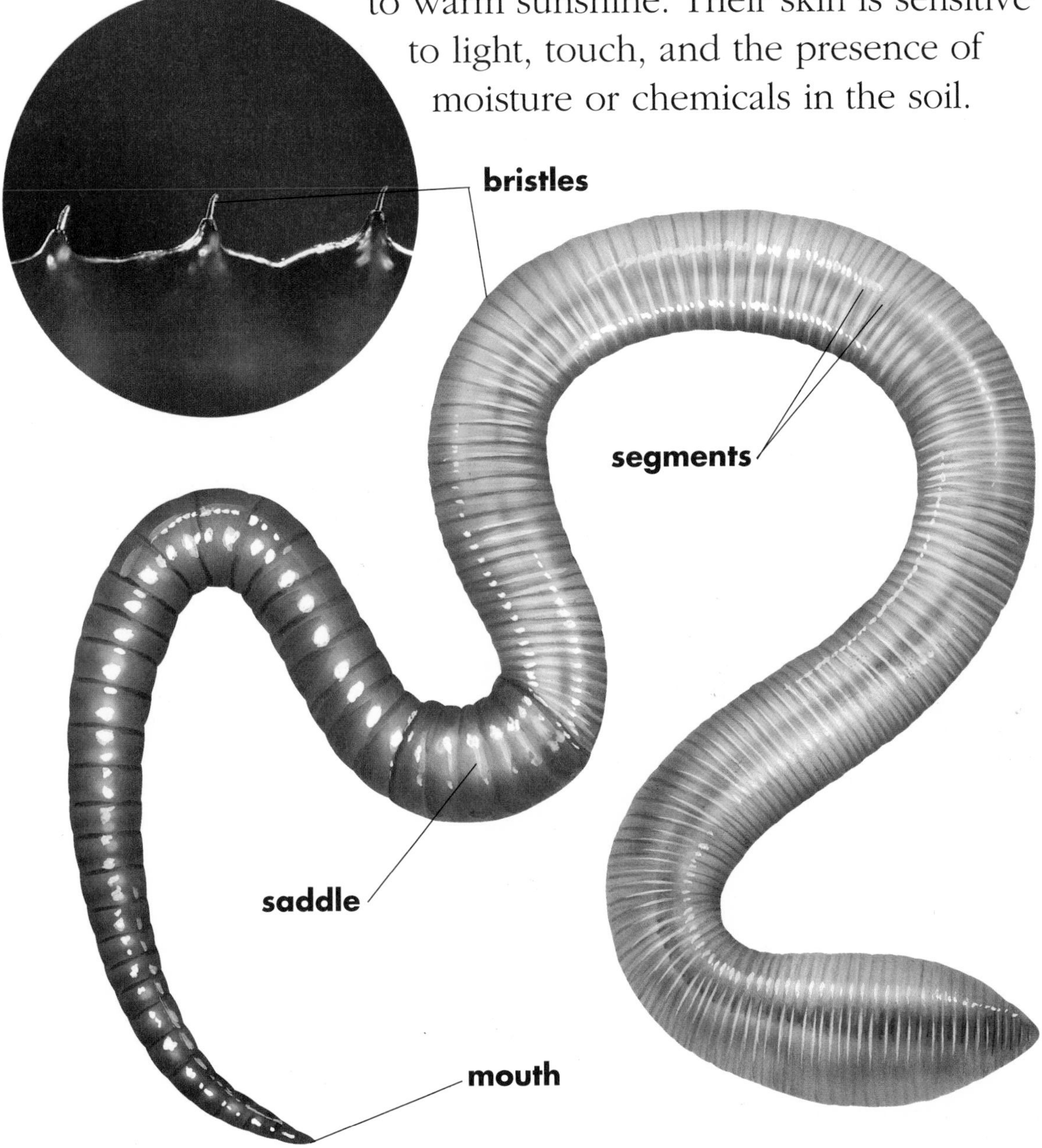

bristles

segments

saddle

mouth

Finding Worms

In mild weather, almost all worms live near the top of the soil. They burrow through the earth finding things to eat. Worms eat bits of dead plants and the bacteria that live in soil. They like rich, fertile soil, because it is full of this kind of food. Small worms often live among the moist litter of fallen leaves in wooded areas.

Worms that swim

Bristle worms are relatives of earthworms but they live in the ocean. Although it is beautiful, this bristle worm is poisonous.

Worms like damp soil, but most of them do not like soil that is always too wet. On a rainy night, worms often leave their burrows. They crawl around on the surface looking for fallen leaves.

Look under stones, flowerpots, or logs. You will also find plenty of worms in **compost** piles.

▼ Bright red sludge worms like smelly pond water! They thrive on the mud at the bottom of the pond. These worms have red blood, which shows through their thin skin.

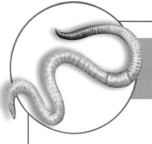

Worm Collecting

First find some jars or plastic containers to put your worms in. If you do not want to touch them with your fingers, take a big plastic scoop with you. Remember that worms have soft bodies, so you must handle them gently. You will also need a trowel or a small spade.

Look under stones, damp logs, and in **leaf litter**. Carefully dig into the soil with your spade until you uncover a worm. Dig slowly, or you might slice one in half! Pick it up carefully, and put it into a jar with a little bit of soil. Write on a label where you found the worm and how deep it was burrowed. Put a lid with airholes on the jar.

Make sure you put your containers in the shade
if you are searching for more worms. They do
not like to get too warm.

Heaps of worms

If you have a compost pile in your backyard, carefully turn some of it over.
Sometimes there are hundreds writhing around! These are called brandling
worms. Scoop some into a plastic container with a spoonful of wet compost.

Identifying Worms

Most worms are known simply as "earthworms," although there are many different kinds. If you collect a lot of worms, you may notice that they are different colors. Some may be much bigger than others.

To find out how many different kinds of worms you have, you will need to write down in a notebook where each one was found, what kind of soil it was living in, and its color. Using a plastic ruler, try to measure the length of each worm.

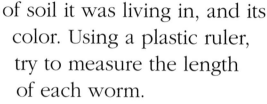

A book on the wildlife of your area or a natural history museum may be able to help you identify your worms. Don't worry if you are not able to tell one from another. It is quite difficult!

Marine worms are segmented, just like earthworms. But they often have lots of long bristles that are easy to see. Some are covered in little colored scales, and some look almost hairy, so it is not easy to see the segments. They do not look like their earthworm relatives.

▲ Some worms, such as the bright green paddle worm, are very pretty. The paddle worm lives in tide pools at the edge of the ocean.

Giant worms

The biggest earthworm comes from Australia. The giant earthworm can grow to over 10 feet (3 m) long!

Homes for Worms

Although most worms look very similar, different kinds prefer different homes. Make homes that are as similar as possible to the places where you found your worms. If you found them on a lawn, dig up some small tufts of grass to put in their new home. Remember that all worms spend most of their time burrowing underground.

Alternate layers of crumbly soil, leaf litter, or compost with a layer of sand about 8 inches (25 cm) deep in a narrow glass or plastic jar. Spread some damp, fallen leaves over the surface.

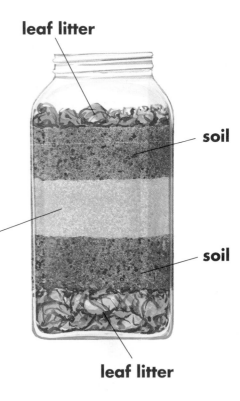

leaf litter

soil

sand

soil

leaf litter

Sprinkle some water over the leaves and soil. Then place your worms on top. Now watch how they burrow!

Cover your wormery with black paper or cloth, or keep it inside a dark box. Worms stay hidden most of the time. If they come out during the day, there is a danger that they will dry up and die.

If you keep your worms in a clear container, you may be able to see the tunnels that they make through the soil.

A house of sand

A marine worm called the sand mason makes its home by cementing sand and tiny fragments of shell onto sticky mucus. The mucus oozes out of its skin.

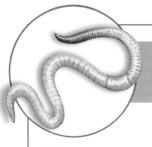

Looking after Worms

Worms do not like their homes to get too hot, too dry, or too wet. As you collect your worms, feel the soil they are in. It's probably crumbly and moist. Try to keep the soil in your worm's home the same.

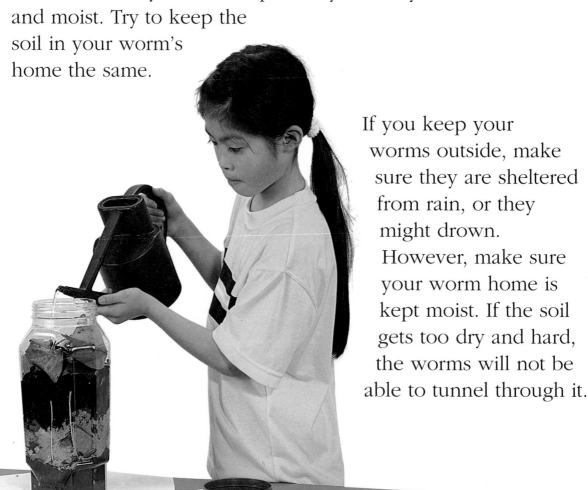

If you keep your worms outside, make sure they are sheltered from rain, or they might drown. However, make sure your worm home is kept moist. If the soil gets too dry and hard, the worms will not be able to tunnel through it.

You will probably find lots of worms living close together in your backyard. Although worms do not fight, do not put too many in each container. They will become too crowded.

There are many creatures that love to eat worms. Birds, foxes, shrews, moles, skunks, and raccoons all eat worms.

Worms in danger

Birds can hear worms moving along their burrows. If a bird pecks the head or tail off an earthworm, it can grow a new one. But worms that are cut in half cannot make two new worms—both halves die.

Feeding Worms

Most worms eat the bits of dead leaves and plants that are found in soil. They also feed on the tiny **fungi**, bacteria, and other creatures that live in the soil. The worm swallows mud as it tunnels. All these bits of plants and creatures are digested as the mud goes through the worm's belly. The leftover mud squeezes out of the back end of the worm like toothpaste.

Some earthworms squirt the mud they swallow into a little pile on the surface of the ground. These piles are called **worm casts**.

Worms often pull dead leaves into the entrances of their burrows. Then they can feed in safety underground. Worms can even pull small twigs and feathers down into the soil. This helps protect them from the probing beaks of birds. The evening is a good time to see this happen. Check every morning to see if any leaves have been pulled down into your worm homes.

▶ If you are lucky, you might catch a worm pulling a leaf into its burrow.

Meat-eaters

Many worms eat the small fragments of dead animals that are in the soil they swallow. But some kinds, like this chaetogaster, eat other tiny **invertebrates** or their eggs.

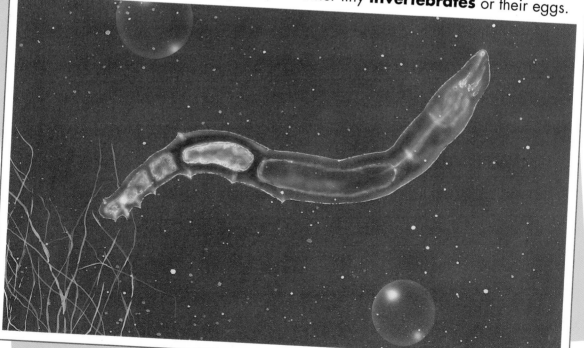

Worm-Watching

Watch how your worms move. Because they have layers of muscles just under their skin, worms can stretch out until they are very long and thin. They can also squash up until they are short and fat. Worms can even be thin in one part of their bodies and fat in another part at the same time.

Pick up a large worm, and hold it in your hand. It is so strong you can feel its snout pushing between your fingers! If you stroke the worm from back to front along the lower half of its body, you will feel tiny bristles.

There are four pairs of bristles on the lower half of every segment of a worm's body. These help the worm grip onto the soil to try to keep birds from pulling it out. A worm's bristles also help it shoot back into its burrow if danger threatens.

▶ There is no need to be nervous about holding a worm.

Moving

Worms move along their tunnels by stretching the front part of their bodies. Then they grip the sides with their tiny bristles and pull the back part.

Multiplying Worms

Some worms mate underground, but others have to come out of their burrows to find a mate. Each worm is both male and female, but they still have to pair up in order to make eggs.

A grown-up worm's body has a thick band around it called the **saddle**. After a pair of worms have mated, they separate. The saddle of each worm makes a collarlike ring where it lays eggs. Then the worm wiggles out of this ring, which seals up and hardens like a **cocoon**.

Hatching

Earthworms bury their cocoons in the soil. Although there are many eggs in each, only one or two wormlings usually emerge. The other eggs provide nourishment for them while they are in the cocoon.

There may be up to 20 eggs in each cocoon. These take between 1 and 5 months to hatch. Each hatchling is like a miniature adult. It has the same number of segments in its tiny body.

▼ A wet evening is the best time to see earthworms mating.

You may be able to watch worms trying to find their mates. Go into your backyard on a mild, damp evening. Worms like darkness, so tape some red cellophane over a flashlight. Walk softly, because worms are very sensitive to vibrations.

Hot and Cold

When the top layer of soil starts to get too cold in the winter, or too warm and dry in the summer, worms tunnel down into the earth. This is because extreme heat or cold cannot reach deep into the soil. Worms may go 3 feet (1 m) below the surface. Some coil up at the bottom of their tunnels and rest.

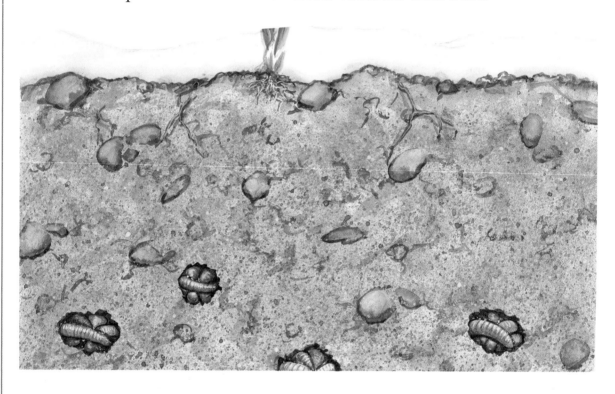

Earthworms have no lungs and no noses, but they do need to breathe. When they are deep in the ground, they absorb oxygen through their skin directly into their blood.

► Lugworms stay cool below the sand when the tide is out.

A lugworm is a kind of bristle worm that lives on muddy or sandy beaches. It makes a large, U-shaped burrow that stretches 10 inches (25 cm) down into the wet sand. When the tide goes out, it stays right at the bottom of the burrow. Some bristle worms swim in the shallow water of tide pools and hide under stones to keep cool.

Water filters

Feather-duster worms live buried in chalky tubes attached to rocks or shipwrecks. They stick a crown of **gills** (breathing organs) out into the water. These sweep microscopic **plankton**, tiny bits of seaweed, and sea creatures into their mouths.

Keeping a Record

You can make a scrapbook about your worms. Record the date you collected them and where you found them. Every time you notice something interesting about your worms, make notes and drawings. If you find pictures of worms in magazines, paste them in your scrapbook, too.

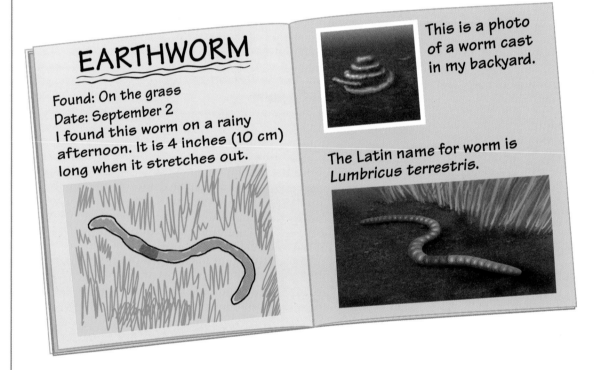

EARTHWORM

Found: On the grass
Date: September 2
I found this worm on a rainy afternoon. It is 4 inches (10 cm) long when it stretches out.

This is a photo of a worm cast in my backyard.

The Latin name for worm is Lumbricus terrestris.

Try to find out more about your worms from books or computer programs. Reading about worms will help you understand how they live. You may even notice something no one else has discovered.

Worms may seem small, but they are very important to farmers and gardeners. Their tunnels allow air to reach the roots of plants.

Worms are constantly eating mud in one place and squirting it out in another. This really mixes the soil. The activity of thousands and thousands of worms churning up the soil helps plants to grow.

Small, but useful

The famous scientist Charles Darwin estimated that the burrowing activities of some worms resulted in about 10 tons of worm casts on the surface of just 1 acre (0.5 hectares) of meadow every year!

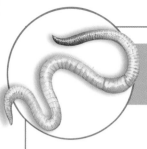

Letting Them Go

Worms can live up to 10 years in captivity, where they are safe from enemies. In the wild, their lives are usually much shorter. This is because so many birds and other animals love to eat them.

When you let them go, take each worm back to the place you found it. On a mild, damp evening, gently tip your worms out of the container onto the ground. Make sure they are hidden by leaves. If not, carefully place a piece of wood over them so they can burrow into the ground safely.

The lugworms and other bristle worms that you find by the ocean are difficult to keep for more than a few hours. They need lots of fresh seawater. But a few hours is long enough to look closely at them, especially if you have a hand lens.

Put the worms into a shallow pan of seawater to watch them. Then pour them back where you found them.

The great escape

Moles are one of the earthworm's worst enemies. If moles find more worms than they can eat, they bite off their heads to keep them from wiggling away. Then they save them for later. If left too long, the worms can grow new heads and escape.

Amazing Worms

Christmas tree worms live inside small, chalky tubes attached to corals. When they feed, they extend pairs of tentacles that draw seawater toward them. The tentacles look very much like tiny, brightly colored evergreen trees.

Honeycomb worms live in groups on rocks or reefs close to the shore. Each worm builds a tube by cementing grains of sand together. These tubes make a honeycomb pattern all over the rocks.

▲ Christmas tree worm

The peacock worm makes a tube by sticking particles of sand and mud together with mucus. The tube is up to 10 inches (25 cm) long. When it feeds, the worm sticks a fan of gills out into the ocean. These draw bits of food down into its mouth, which is in the center of the fan. The gills also help the worm breathe by absorbing oxygen from the seawater.

▶ Peacock worm

Fireworms live right inside the sponges on which they feed. When they come out to breed, the females swim to the surface and make a glowing light. Males make flashing lights. Then they swim toward the steadily glowing females.

The palolo worm lives inside coral reefs around the Pacific islands of Samoa and Fiji. At the same time every year, many thousands of palolo worms swarm to the surface of the ocean to breed. At this time, native people catch many of these worms to eat.

Further Reading

Fowler, Alan. *It Could Still Be a Worm.* Children's Press, 1997.

Glaser, Linda. *Wonderful Worms.* Millbrook, 1992.

Morgan, Sally. *Butterflies, Bugs, and Worms.* Larousse Kingfisher Chambers, 1996.

Moyle, Philippa, and Louise Morley. *Minibeasts* (Pocket Gems series). Barrons Juveniles, 1997.

Ross, Michael Elsohn. *Wormology.* Lerner, 1996.

Glossary

Cocoon An outer case that protects the delicate eggs of earthworms.

Compost The rich, dark litter of decayed plants, vegetable peels, grass cuttings, and so on. It is obtained when all these items are piled up and left to rot outside.

Fungi A large group of living organisms, many of which produce mushrooms and toadstools.

Gills Very thin-skinned organs that help many water animals breathe. They absorb dissolved oxygen from water.

Invertebrates Animals that do not have a spine or backbone.

Leaf litter A layer of fallen leaves, mostly from trees.

Mucus A slimy liquid produced over the surface of a worm's skin to keep it from drying up.

Plankton Tiny plants and animals that swim or float near the surface of the ocean.

Saddle A band of thick skin on an earthworm that oozes out a special thick collar of mucus.

Segments The many ringlike parts that make up an earthworm's body.

Worm casts Leftover mud that has been swallowed and digested by an earthworm.

Index

The publishers would like to thank the following for their permission to reproduce photographs:
cover· (worms) Kathie Atkinson/Oxford Scientific Films, 4 Johan De Meester/Oxford Scientific Films, 5 David
Thompson/Oxford Scientific Films, 6 Michael Glover/Bruce Coleman, 7 G. I. Bernard/Oxford Scientific Films, 9 Kim
Taylor/Bruce Coleman, 11 top G. I. Bernard/Oxford Scientific Films, 11 bottom A.N.T/Natural History Photographic Agency,
13 Rodger Jackman/Oxford Scientific Films, 15 Kim Taylor/Bruce Coleman, 17 Kathie Atkinson/Oxford Scientific Films,
20 David Thompson/Oxford Scientific Films, 21 David T. Grewock/Frank Lane Picture Agency, 23 top Eckart Pott/Bruce
Coleman, 23 bottom Paul Kay/Oxford Scientific Films, 27 David Thompson/Oxford Scientific Films, 28 Ian Cartwright/Frank
Lane Picture Agency, 29 D. P. Wilson/Frank Lane Picture Agency.